All Einstein,
All Physics,
All Controversy

Lonnie Hicks

ISBN: 978-1500104078

ISBN-10: 1500104078

Foreword

This book is a collection of essays done over several years relating to modern physics, physics theory and controversies which surround physics and allied sciences. I have included over 300 links to videos on these topics and many, if not most are related to the ideas of Einstein and others working in various fields.

My goal is to highlight some of the basic ideas in the discipline in a spirit of understanding them from a lay point of view for myself and other non-physicists.

It has been a fascinating journey, one which can and will continue, as many of these essays are on-going.

Readers can get updates on these in the future from my website. The conversation is continuing.

Einstein and Quantum Levitation

This essay explores the possible relationship between Einstein's physics and Quantum Levitation.

Thereafter I move to examining the role of magnetism and superconductivity. The format I have chosen is one which looks at developments in physics on a daily basis such that the reader can get a sense of real time developments. It is also useful, I hope to researchers who want to get a sense of how various ideas developed.

So why, you ask, should you care about this?

Contents

Mathis

Updated: 8/2/12 Does the Higgs exist and it is that we simply have not recognized it as such?

Updated: 8/1/12 Does the Nature of the Universe and the Higgs depend upon the nature of the Photon?

Updated: 7/31/12 The Mathis View of the Higgs

Updated: 7/30/12 Quantum Gravity and the Higgs. My head just exploded.

Updated: 7/29/12 Deducing what the Higgs field might be.

Updated: 7/28/12 Spintronics, Gravity and Einstein

Updated: 7/26/12 Is Accelerating Iron ions at CERN looking for the HIGGS reckless? See below

Updated: 7/23/12 Empty Space as a Superconductor?

Updated: 7/21/12 Cold Gases and Hot Superconductors

Updated: 11-9-11 So was Einstein Right or Wrong? Answer: Wrong

This blog is a continuation of my explorations of modern physics theory and specifically this time I want to explore the possible relationship between Einstein's physics and quantum mechanics.

Today I want to start the exploration of quantum levitation and Einstein's famous equation E=mc2.

Exactly what is that you ask?

I am not sure yet, but we shall see.

But first read the article below on superconductivity and the amazing You Tube video which demonstrates quantum levitation. Then we come back and start to explore the implications of this experiment and Einstein's physics.

Prepare to be astonished by this video.

http://www.youtube.com/watch?v=Ws6AAhTw7R A

http://www.youtube.com/watch?v=Woc9wbGF_lQ &feature-related

http://en.wikipedia.org/wiki/Superconductivity

Tomorow we're back to have a look at what this might mean.

 See the two videos below on superconductivity and how levitation works then we will delve into the possible meaning of all of this.

http://www.youtube.com/watch?v=fPV2QsNasT8 &feature=related

11-6-11

Now what does this levitation mean? There are
several possible implications, ones which are not
discussed by the main stream science
establishment.

First this levitation is possible because electrons
and ions are being held in place under
superconductivity conditions by a magnetic field
and magnetic lines of force. They theoretically
could continue to move in this frictionless fashion
forever.

It is a possible source of incredible energy if made
practical. But the problem is that in order to reach
super conductive states materials have to be super-
cooled to near absolute zero, although some
materials do not have to be cooled that much.

But the process takes time and incredible amounts
of energy to do the cooling in a process which is
very expensive and uses a lot of energy in the
process.

Secondly moving electrons, protons and ions the
like inside a super-cooled magnetic field is exactly

the same process being used at CERN to smash atoms, a project I have questions about.

Thirdly, I have also suggested that this is the process at work in our universe at large. See my articles on the Plasma Theory of the Universe on this site.

So we see the relativity theory, Plasma Theory and the Quantum world are now all at least employing the same techniques in examining the world while not acknowledging what the implications of all this are.

Here the implications:
If nature at the quantum level operates on magnetic lines of force operating on sub-atomic particles, and if the entire universe including galaxies and stars also operate in the same way, then these elemental forces are so powerful that one has to question messing with them at CERN.

It means that the Standard Model has some serious flaws and likely that the giant magnets utilized there at CERN could affect the magnetic field of the earth itself and the planet could be extinguished--especially if the Plasma Theory of the Universe is correct.

(See my blog on the Plasma Theory of the Universe.)

It means that our universe is a dangerous place and we have underestimated the danger and possibly have exacerbated those dangers with the CERN experiments.

It means that modern physics in searching for the Higgs field and the Higgs Boson at CERN is, in effect, acknowledging the existence of Aether and that modern physics took a wrong turn to particle physics and should have kept to the course as regards understanding the nature of space, not particles.

Most of the Universe and the sub-atomic world is space not particles and physicists know very little about it assuming for over a century that space was "empty."

This turns out not to be true.

Space is alive with a field we can't detect and have named this empty space "Dark Matter" and "Dark Energy" when in fact space is just like the levitation experiment identified above and we, and the planet earth, are the hockey puck held captive in these tremendously powerful magnetic lines of force in gigantic magnetic fields which dominate our galaxy and our universe. These gigantic magnetic fields in our galaxy are propelling ions in orbits millions of light years long-just like we use giant magnetics to propel ions and photons at CERN. Both are identical processes and this same process is at work in the levitation experiment above, in my view.

(See my blog on this on this site as well."

Now these huge magnetic bubbles, (4) (which are the largest structures in our galaxy,-over 50,000 light years across in our galaxy which is only 100,000 light years across) are clearly the drivers in the galaxy if not the universe.

Astronomers ignore these magnetic bubbles and their possible function in the galaxy. But to me it is clear space is magnetic which means magnetics not gravity is what holds the universe and the hockey puck too, in place.

Hannes Alfven, wherever he is, is smiling.

But how does all this relate to Einstein's theory of relativity? We shall see tomorrow.

11-9-11
Now you ask how does all this relate to Einstein's theories.

Well first let's just review a little.

The part his theory I am relating to is, for our

purposes, the proposition that all of space-time is a fabric-grid and can bend and is affected by mass or bodies. We want to know how is all this possible.

Einstein's only answer was gravity, a notion of gravity which is essential to Newton notion of gravity transformed in a space-time fabric. But since space-time is not defined but merely described we want to know what is there about space-time which gives it the character of a fabric?

Ignoring gravity for the moment let's just focus on space-time. If space-time is malleable then we ask what is space time like in a black hole where space-time also is affected by gravity to an incredible degree; is space-time like a Higgs field where some particles pass through it is affected, where the length of an object is shortened as the object approaches the speed of light; is it like a mass which grows larger and larger as the object approaches the speed of light, is space-time itself relative to the positions of the observed and the observer?

Lots of questions, but the central question we are left with is what is space-time?

Telling me how it behaves is not the same as telling me what is, and more over the question of how it relates to the sub-atomic world is never answered. Einstein didn't have answer and spent the last 30 years of his life looking for one.

But note Einstein had a bias; he was writing his theories in the beginning looking to refute notions of Aether which had dominated in science for over 150 years. (See my blog on Einstein to get details.)

Here I propose an answer to the question of what is space-time since Einstein did not.

Here goes.

Space-time bends because it is a particular form of Aether where gravity, dark energy, dark matter and magnetic field theory converge.

To see bending we must first see the rigidity which underlies space-time and that rigidity is just like the Quantum levitation described above. Magnetic fields hold the universe together and that field is similar to the ones we know on earth except they operate on whole galaxies as well as on the quantum level.

Now we want to know what happens to these magnetic fields such that they bend.

Here is my idea.

As a body or even electrons move through the medium of empty space electrons actually move into existence and fall out of existence continuously much like tiny waves appear and disappear on the surface of this bendable ocean of existence. That is why electrons can behave both as a particle and as a wave.

See: http://www.sciencedaily.com/releases/2011/11/111118133050.htm

It gives this magnetic ocean ground of being its bendable character since existence itself can be created, refracted, reflected yet the ocean retains it's over--all character. It is still an ocean but is alive, constantly changi

The magnetic field plasma holds this ocean together and forms the basis of all existence in place very much like quantum-trapping on the sub-atomic level, and huge magnetic fields on the galaxy level.

Gravity is involved but gravity is the capacity of ions, electrons and the like to instantly re-align their polarities in the presence of larger bodies and masses. Some call this quantum pairing without understanding its implications or relating it to other phenonmena.

Electrons can jump their orbits, ions can be created by these sudden changes in polarities and their realignment is essentially due to magnetic forces and magnetic fields.

Our solar system, our universe, and the sub-atomic level all are held in place and bend due to shifting polarities of objects which exist and operate in the field.

There is a Higgs field, except it is a magnetic field which governs all of existence.

Electricity for example can be bent and flows of electricity seek equilibrium. Where that equilibrium is disturbed events occur, some good, some bad.

So, what about Einstein? What Einstein missed, in my view, is that space-time is being bent by the magnetic polarity oscillations which exist between two bodies. Bending is shifting polarities which occur, along with the fact that our moon and our sun are held in place orbiting around each other not by gravity but by exactly the magnetic lines of force we see in these videos on levitation.

But levitation is not the critical fact.

The critical fact is that the hockey puck is being held in place by magnetic properties of objects under superconductive conditions.

This explanation also explains quantum mechanics weirdest contention--that the observer affects the behavior of electrons. Why? Because the observer, scientist is a source of polarity shifts between himself or herself and the electron behavior being observed. The observer is made of polarity shifting electrons which affects what is being observed.

All of space then exists in a field which has a form of superconductivity which has to be explored.

Here comes Aether back around again.

Now this theory is not new and Hannes Alfven actually proposed it several decades ago. (See the blog on this site on Plasma Theory)

See also Alfven waves in the Einstein blog for details.

The implications of this view if true are enormous and spelled out there.

Now, in the simple example of the earth and sun, gravity or magnetic lines of force are at work here as well, not relativity or special relativity because the action of sun on the earth is instantaneous, and light is much slower than that.
Einstein's "spooky action at a distance" is actually this Plasma aspect at work and it is instantaneous.

Space-time operates on light but no such speed limit apparently operates on electrons traveling in space-time described in terms of magnetic lines of force moving under superconductivity like conditions.

We see this also operating in the video but no one seems to understand the implications of what the experiment may mean.

So instant impact, not slow light speeds, is what governs the universe rightly understood as being held together and governed by magnetics and huge plasma fields on both the sub-atomic level and the galaxy level.

Astronomers are beginning to recognize that magnetic fields really dominate the universe. See this Sciencc Channel episode on "Mega flares" Alfven ideas are beginning to be recognized as well has the dangers this facts means for Earth.

http://science.discovery.com/tv-schedules/series.html?paid=48.17206.133485.36228.x

Now am I correct in all of this? How could we experimentally prove the Plasma theory?

That next. Meantime:

http://www.nist.gov/pml/div689/jila_070610.cfm

http://www.sciencedaily.com/releases/2008/08/080801153127.htm

http://www.sciencedaily.com/releases/2011/02/110223151943.htm

Refresher on the Higgs Field before the plunge on
.
http://profmattstrassler.com/articles-and-posts/the-higgs-particle/360-2/

7/23/12

Now we face the question squarely. Can empty space operate in the context of magnetism and superconductivity?

This fellow thinks so.

http://en.wikipedia.org/wiki/Maxim_Chernodub

Observations on superfluids in neutron stars

http://www.sciencedaily.com/releases/2011/02/110223151943.htm

Quote from the article above:

"Chandra data found a rapid decline in the temperature of the ultra-dense neutron star that remained after the supernova, showing that it had cooled by about four percent over a 10-year period.

"This drop in temperature, although it sounds small, was really dramatic and surprising to see," said Dany Page of the National Autonomous University in Mexico, leader of a team with a paper published in the February 25, 2011 issue of the journal *Physical Review Letters*. "This means that something unusual is happening within this neutron star."

Superfluids containing charged particles are also superconductors, meaning they act as perfect electrical conductors and never lose energy. The new results strongly suggest that the remaining protons in the star's core are in a superfluid state and, because they carry a charge, also form a superconductor.

"The rapid cooling in Cas A's neutron star, seen with Chandra, is the first direct evidence that the cores of these neutron stars are, in fact, made of superfluid and superconducting material," said Peter Shternin of the Ioffe Institute in St Petersburg, Russia, leader of a team with a paper accepted in the journal *Monthly Notices of the Royal Astronomical Society*."

Now add this to observations that the core of our

very sun might have some similar characteristics and we have a tantalizing mystery especially if we also observe that superfluity and superconducting inside and between huge galactic magnetic fields is possible.

Here are a few links:

http://www.sciencedaily.com/releases/2011/08/110801094253.htm

http://www.sciencedaily.com/releases/2011/11/111123133137.htm

http://www.sciencedaily.com/releases/2012/07/120709092457.htm

"If these motions are indeed that slow in the Sun, then the most widely accepted theory concerning the generation of solar magnetic field is broken, leaving us with no compelling theory to explain its generation of magnetic fields and the need to overhaul our understanding of the physics of the Sun's interior."

Quote from the above article.

This is getting very interesting.

http://www.thesurfaceofthesun.com/model.htm

Supernovas and iron

"The researchers found clumps of almost pure iron, indicating that this material must have been produced by nuclear reactions near the center of the pre-supernova star, where the neutron star was formed."

Some argue that our sun was created from a supernova and has a core of iron. Wow. Here is a link because if true then our sun's magnetic field is being generated from such a hypothesized iron core.

Gulp.

http://www.thesurfaceofthesun.com/
http://www.thesunisiron.com/archives/Conference_on_Physics_Beyond_the_Standard_Model-Finland.pdf

http://www.sciencedaily.com/releases/2012/03/120329124724.htm

http://philosophyofscienceportal.blogspot.com/2008/03/iron-core-for-sun-and-more.html

http://www.spaceref.com/news/viewpr.html?pid=17308

The moon as dynamo
http://www.sciencedaily.com/releases/2011/11/111
109131821.htm

What does all this mean? Well possibility that
these bodies relate to one another across and
within generated magnetic fields

So where is the matter located in "empty space"
and its distribution?

http://www.sciencedaily.com/releases/2008/05/080
520152013.htm

7/26/12

Now if the stars and our very sun have iron
components and we know that iron is the cause of
supernova explosions we should examine this iron
connection closely.

Now we understand that CERN is going to double
the speed of its particle accelerations this time
using iron and lead ions, common sense asks if this
is wise-especially the iron.

http://www.rmki.kfki.hu/~plevai/QGP/ann_abc.ht
ml

Quote below from article.

"Scientists at CERN say a series of experiments slamming iron ions together provide strong evidence that they have recreated this primordial state of matter."

Iron is dangerous for the stars and their stability, resulting in the greatest explosions known to man, we ask the simple question what is the danger at CERN?

Critics say the iron ion experiments are foolhardy and dangerous.

Let's let the critics speak for themselves. See the link below:

and come back and we discuss.

http://www.authorsden.com/visit/viewshortstory.asp?AuthorID=121255&id=48865

See also: Iron is abundant in the Earth's core and help protect us from the Sun's deadly rays. Does the core conduct electricity. Yes, say these scientists

http://www.sciencedaily.com/releases/2011/12/111219112216.htm

Quote:

"Compounds typically undergo structural,

chemical, electronic, and other changes under these extremes. Contrary to previous thought, the iron oxide went from an insulating (non-electrical conducting) state to become a highly conducting metal at 690,000 atmospheres and 3000°F, but without a change to its structure."

http://www.sciencedaily.com/releases/2012/05/120508094346.htm

"In order for a dying star to end up as a supernova, its core must have been transformed into iron."

"Once the core cannot be compressed any further, the compressed matter must expand again in a gigantic explosion, or supernova. This is where the heavy elements of the universe may have been formed."

More:
http://www.sciencedaily.com/releases/2012/06/120625125954.htm

http://www.sciencedaily.com/releases/2008/05/080528140242.htm

http://www.sciencedaily.com/releases/2010/09/100910101834.htm

http://www.youtube.com/watch?v=tr8qvwJQkbM&feature=youtube_gdata_player

So iron matters and we need to have a look at what its role and dangers might be in the LHC.

That next.

7/27/12 Iron's the thing

But first a discovery that a single iron substance can be both magnetically and electrically polarized, unheard of--at temperatures close to absolute zero. And what may we ask is the temperature of space? It is a few degrees above absolute zero. And what is the temperature of the liquid helium at CERN being used for cooling. Near zero.

http://www.sciencedaily.com/releases/2012/06/120624134945.htm

Quotes from the article:

"Now researchers at the Niels Bohr Institute at the University of Copenhagen have studied a material that is simultaneously magnetically and electrically polarizable."

I would add "at the same time."

Another quote from the article.

"It is precisely this interaction between the

transition metal, iron, and the rare element, terbium that plays an important role in this magneto-electrical material. The terbium's waves of spin cause a significant increase in the electric polarization and the interaction between the ions of the elements creates one of the strongest magneto-electrical effects observed in materials."

Terbium?

http://www.lenntech.com/periodic/elements/tb.htm

Now interestingly this was predicted by Alfven who urged that space is not just magnetized clouds of dust but magnetized, polarized vessels which also carry polarized electric charges and flows. These are not exclusive of one another. The article above makes the additional point that such configurations can also involve polarization, a point I strive to make above.

Now thinking about the earth's core as such a vessel we have iron with the ability to become both polarized both magnetically (we are familiar with that in magnets) but also electrically. Electron spin is polarized spin and can happen in iron.

If it does, uncontrolled as by the magnetics at CERN, then the core of the earth, iron, can in fact gain electrical capability and become unstable.

An ugly thought.

So who is thinking about this at CERN? Who is doing something to ensure it does not happen. After all, what has been built as CERN is the largest magnets in the history of the world.

So just maybe somebody ought to be clearly spelling out these dangers and give the public some assurance that the dangers do not exist; because if they are wrong there will be no "do over."

7/28/12

Next we want to ask how this dual polarization process might work in interstellar context and ally this to the ideas around spintronics.

Above I introduced the idea that electron spin and polarization changes at the quantum level exist, are real, and might play a role in gravity, not just magnetism and electricity. I have argued that similar processes operate at the macro level and the micro level. Let's examine the evidence around this.

Here is a background article on electron spin after that we will go back to the question on the floor: can the universe operate on low-energy electron spin, electricity-magnetics?

http://www.sciencedaily.com/releases/2007/11/071101144942.htm

http://www.sciencedaily.com/releases/2009/06/090617080717.htm

Materials and electron spin: common examples

" $Cd_2Os_2O_7$ has the intriguing property that when cooled to 227K (-46 °C), it undergoes both a metal-insulator transition and a magnetic transition to a state in which all its electron spins are aligned. This spin alignment, which makes the material magnetic..."

Spintronics comes in many varieties. See quote from article below:

http://www.sciencedaily.com/releases/2012/07/120703181909.htm

"The groundwork for Lufkin's breakthrough was laid several years ago, when researchers discovered that nitrogen-vacancy (NV) centers, atomic-scale impurities in lab-grown diamonds, behave in the same way as single atoms. Like individual atoms, each center possesses a spin, which can be polarized, similar to on a bar magnet. Using lasers, researchers are able not only to control the spin, but to detect its orientation as it changes over time."

On the different kinds of magnetism and spins.

http://en.wikipedia.org/wiki/Ferromagnetism

http://en.wikipedia.org/wiki/Spin_(physics)

http://en.wikipedia.org/wiki/Magnetic_dipole_moment

http://en.wikipedia.org/wiki/Ferromagnetism

http://lofi.forum.physorg.com/Cause-Of-Magnetism_20934.html

Next time we sum up together the various threads we have alluded to above and ask can we get a coherent theory of the universe out of it one that is based on experimental data and actual observations?

Tall order but we'll see.

7/29/12

More critics and then we move to the possible description of the Higgs

http://www.scienceguardian.com/blog/kicking-mother-natures-shins-lhc-renewed-startup-proceeds-towards-biggest-bang-ever.htm

Oxford Scholars assess the risks of CERN

Weighing in on CERN's safety are scholars from Oxford. See link below and then we discuss. See what you think.

Click on the author's name (Toby Ord) in blue in the article to get a PDF of the article.

http://www.newscientist.com/article/mg20126926.800-how-do-we-know-the-lhc-really-is-safe.html

A quote from the above paper after examining CERN's analysis of the safety of the project the authors goes on to say:

"While the arguments for the safety of the LHC are commendable for their thoroughness, they are not infallible. Although the report considered several possible physical theories, it is eminently possible that these are all inadequate representations of the underlying physical reality. It is also possible that the models of processes in the LHC or the astronomical processes appealed to in the cosmic ray argument are flawed in an important way. Finally, it is possible that there is a calculation error in the report."

More:

"However, our analysis implies that the current safety report should not be the final word in the safety assessment of the LHC."

"To proceed with the LHC on the arguments of the most recent safety report alone, we would require further work on estimating $P(\neg A)$, $P(X|\neg A)$, the acceptable expected death toll, and the value of 15 future generations and other life on earth. Such work would require expertise beyond theoretical physics, and an interdisciplinary group would be essential. If the stakes were lower, then it might make sense for pragmatic concerns to sweep aside this extra level of risk analysis, but the stakes are astronomically large, and so further analysis is critical."

A critique of the science behind CERN

"For a devastating critique of hundreds of years of Physics math see the link below. Note especially the theory of gravity examined shows gravity is meaningless without including the electromagnetic fields involved."

I do that below in my ideas the subject (independently.)

http://www.milesmathis.com/central.html

another quote from the above article:

"Relativity theory is a miniscule part of modern physics. Very few people know anything about it. The few that do are working on billion-dollar projects—to discover the graviton or launch the next satellite. The last thing they want is some theoretical controversy to get in the way of funding."

"I began this book when I stumbled across the first great error many years ago, in reading Einstein's *Relativity*. Although it soon became apparent that the error was both elementary and profound, I thought at the time that it was an isolated error. But my naiveté evaporated as I subsequently reread other important theoretical papers, and my awe of the past evaporated with it. What I came to realize, with rising disbelief (as well as some excitement), is that my faith—the faith of all scientists—in the basic theory and math of physics has been unfounded. It became apparent that the theory and math of many famous and influential papers, both classical and modern, had never been checked closely—or not closely enough for my taste at any rate. Buried in these papers were algebraic and geometric errors of the most basic kind. Suffocating beneath dense, often impenetrable theories and unnecessarily difficult equations of so-called higher math were errors that a high school student could understand, were he or she presented with them in a straightforward manner.

My goal became to do just that. To strip physics of its mystifying math, its unnecessary proliferation of variables and abstract concepts, its stilted language and dry jargon, and to speak in clear everyday sentences and simple equations. Einstein is famous for stating that a theorist should be able to explain his theory to an eighth grader, but he did not practice what he preached."

And what is that simplification? Mathis comes to a conclusion I have been arguing for:

There are only two forces: Gravity and the EM force.

"Unifying the two major fields of physics like this must have huge mathematical and theoretical consequences. Because Coulomb's equation is a unified field equation, gravity must have a much larger part to play in quantum mechanics and quantum electrodynamics. Gravity must also move into the field of the strong force, and require a complete overhaul there.

By the same token, the E/M field must invade general relativity, requiring a complete reassessment of the compound forces. At all levels of size, we will find both fields at work, creating a compound field in which each field is in opposition to the other.

Yes, according to my new equations, the two fields are always in vector opposition. And since gravity, by itself, is a function of radius alone, it must be much larger at small scales than we thought--and somewhat smaller at large scales."

http://www.milesmathis.com/uft2.html

See also:

http://www.scienceguardian.com/blog/kicking-mother-natures-shins-lhc-renewed-startup-proceeds-towards-biggest-bang-ever.htm

So today we begin to bring together various actual experimental data and actual observations in an attempt to deduce the characteristics of the Higgs field and place these deductions in the context of a new view of how the universe works. Ambitious, true.

First let's ask some basic questions as to what the nature of the Higgs
field might be.

The Higgs field is likely low energy, not likely to be detected by high energy LHC particle searches. LHC is designed for high energy particle searches and even some of the various HIggs postulated are low mass, low energy possibilities and they will be missed.

This makes the iron-lead ion strategy not only dangerous but, also, likely futile.

Note also that dark energy calculations make it clear that it does not take much energy to make the universe expand, especially if we are talking superfluidity and superconductive conditions, which we have alluded to above. This explains partly as to why such an expansion can accelerate in the presence of postulated Alfven waves as an additional factor.

Looking for high energy particles in the detectors is perhaps the wrong way to go about it and has, instructively, produced little to date.

This again reinforces the notion that the Higgs field is a low energy field not a high energy field.

It doesn't have to be.

Quantum data and superconductivity data suggest that the field will likely be polarized both magnetically and electrically at the same time and can be so at both at the gaseous state and at low temperature superconductor states.

Superconductivity and temperatures in space happens at low kelvins are related.

We want to know how.

But interestingly is also true that high-temperature superconductivity and superfluidity, as in a neutron star, is also possible and has been observed. See above article on this.

So if we can see the universe as predominately plasma, the 4th state of matter, the Higgs would likely be a part of that overall state of affairs-- both on the galaxy and the quantum levels.

Electron spin and polarity changes we have seen can be drastically altered indeed controlled in materials, (and gases too I would guess) in the context of "empty space" being polarized magnetically and electrically at the same time in a field we can call a low energy Higgs for the time being.

This capacity is also a way of understanding that the Higgs field is likely involved in the propagation of gravity, in some sort of polarization process or at least involved in its propagations via wave functions, this at speeds faster than the speed of light. Light takes eight minutes to reach earth. Gravity is instantaneous, obviously much faster. This is gigantic quantum pairing, on a large scale, no?

More, we now know that quantum pairing can happen among all of the quanta not just two pairing. Think on that for a moment, in a gravity context.

So the question becomes not how Higgs creates mass, but how Higgs creates gravity and more ominously, the collapse of gravity and there might be such a thing as an EM collapse process involved as well. Fields can collapse we know that, but how does is work in space in the interplay between gravity and EM fields and the Higgs?

We see gravity collapses in stars and my fear is that we might see it on earth as well-especially when we understand the role of iron on the stellar level and on the quantum level.

We need to examine the above in more detail but more importantly we need to get genuine concerns out to the public and have the CERN people look again at their assumptions.

Newspapers may now become interested in this story and CERN officials would be wise to get out in front of this story and take remedial safety related actions to show responsiveness.

Mercifully, they shut down in Feb 2013 for two years. Two years is all we have to get the facts out and the risks objectively looked at. See my article above by Oxford statisticians who have done the only objective assessment of CERN. Their report has been ignored because it concludes CERN's self-serving assessment of its own safety is, surprise, flawed.

Likely CERN will clam up in the next two years and hope it all goes away.

That would be a bad response because the iron-lead ion plan is fraught with dangers which even lay people like myself can understand. Lead ions also in use may not be much better that iron.

Iron experiments at CERN:

https://search.cern.ch/Pages/results.aspx?k=iron

So if you would forward this blog to those who might understand the issues and let's put publicity on all of this.

My email:

lonnie.hicks.gmail.com

7/30/12

What is dark matter and how does it relate to the Higgs?

But first see what you think of this fellow's ideas-they are extraordinary.

http://milesmathis.com/massless.pdf

On the Higgs announcement July 4th and dark matter/

http://milesmathis.com/higgs.pdf

Gravity as an accelerant and EM as an attractor

http://milesmathis.com/uft.html

"...gravity is not an attraction. It is a motion. It is real acceleration, and it is a real acceleration in the direction that a real acceleration is required to

<u>create the force. That is, its direction is outward from the center of the Earth.</u>

This is a shocking statement. We shall have to come back to it later.

http://milesmathis.com/quantumg.html

Gravity at the quantum level. See what you think of this fellow's idea on quantum gravity.

http://milesmathis.com/catas.pdf

These are extensive quotes from Mathis who whose ideas bear on the Higgs. Read and then we discuss. Forgive the long quote.

"All the waves in the data and math are caused by photons with real spin and radius. It is this radius of the photon and the stacking of spins that causes both the waves and the quantization, as I have shown in
many previous papers.

The only analogue to a field wave we have at the quantum level is the neutrino, which is a field wave in the charge field, moving at c. But that isn't what we are seeing in the wave function or in the normal motions of the quantum charge wind. The waves we are detecting in most quantum experiments are stacked photon spins, not neutrino field waves. Slater and de Broglie were not seeing neutrino field waves in the data; they were seeing real photons spins. They were seeing quanta being pushed by a photon wind.

I said above that I would be able to explain the empty wavefunction, and I will do that now. The empty wavefunction is not really empty; it is just a charge field without any quanta to push around. The charge field is not created by matter. Although it is recycled by what we now call fermions, it exists with or without fermions. Photons do not require the presence of other matter. Charge density is increased by the presence of other matter, since concentrations of matter tend to attract photons."

"But even nearly empty space will have photon traffic, as we know. What we don't seem to know is the baseline density of that traffic. It is far above what we have thought. Because most photons are dark to us, and because we have no way to measure photon traffic in the absence of ions, we have not understood that this baseline is very high. I have shown how to calculate this baseline straight from the fundamental charge, showing that what we call dark matter is actually photonic matter. I first unveiled
this calculation in my paper analyzing MOND:

$e = 1.602 \times 10^{-19}$ C 1C $= 2 \times 10^{-7}$ kg/s (see definition of Ampere to find this number in the mainstream) e = 3.204 x 10-26 kg/s

If we divide that last number by the proton mass, we get 19, which means that charge outweighs baryonic matter by 19 to 1, or 95%. That is the current number for dark matter.

This means that the empty wavefunction is just another measurement of the charge field. It is telling us we have charge with no particle to lead around.

That is now easy to understand, given the charge field, which is real photons. The empty wavefunction is real charge photons with no ions. This is how charge is transmitted across "empty" space at the speed of light. Charge is light."

"My charge photons don't travel via an ether. All the waves in my quantum theory come from the real spins of real particles, not from traveling as waves in a field. In other words, the wave in my quantum theory belongs to each individual photon, and is a real motion of that particle.
And photons spin from their own collisions, not from the influence of an underlying ether. This bypasses all the philosophical quibbling of the 20th century. Because I give the spin to a real particle in a mechanical field—rather than to an undefined math or poorly defined ether—my

theory succeeds where the others failed."

Finally Mathis seems to be saying in the article below there is one particle, the photon and all others are derived from it by spins. Golly

http://milesmathis.com/elecpro.html

So what is being said here? Let's discuss tomorrow.

7/31/12

But first a little bit more on the earth and its iron-related aspects.

http://www.sciencedaily.com/releases/2007/09/070920145537.htm

http://www.sciencedaily.com/releases/2008/11/081112140404.htm

What about Mathis' ideas?

Now what do we have with the Mathis ideas? There are many and we shall wind our way through them and see how compatible they are with the ones already outlined above.

First, what is striking about Mathis and his ideas is that first he takes on the entire math basis of the most important equations in Physics for the last 300 years or so. Whether that enterprise has been successful or not I am not prepared to say.

Here are his conclusions if you want to read ahead.

"In short, the scientists and mathematicians have insisted on inserting physical points into their equations, and these equations are rebelling. Mathematical equations of all kinds cannot absorb physical points. They can express intervals only. The calculus is at root a differential calculus, and zero is not a differential. The reason for all of this is not mystical or esoteric; it is simply the one I have stated above—you cannot assign a number to a point. It is logical and definitional.

http://www.milesmathis.com/central.html

But for now , I am interested in the theoretical ideas which he puts forth.

Today let's look at the one identified above and start the comparison to what I have proposed.

First of note is that both he and I agree that the major forces at work in the universe and at the quantum are essentially just two forces--gravity and the EM force. The strong and weak forces are simply aspects of the latter two.

His demonstrations of that argument are mathematical and I will not attempt to evaluate them just yet but, in the main from, what I have so far read, an important component is that he argues that physicists are poor mathematicians and have misapplied ideas around velocity and confounded simple equations into needlessly complex ones.

Perhaps; no, I agree mostly.

Now the idea we want to explore is what Mathis has to say concerning the Higgs. His view is simple--the Higgs does not exist-if I am correct, or it exists in a different form that currently contemplated.

Rather what exists is photon spins and what he calls "photon stacking."
All of the other artifacts of endless particle building he dismisses as not existent or reducible to photon originations. This is electrons, protons and the like, neutrons and the like.

http://milesmathis.com/elecpro.html

Wow.

It is actually not a new idea. As we can see there is experimental evidence from experiments cited above where scientists have actually observed and manipulated photons into and out of existence. Mathis simply asserts "charge is light." This is extraordinary.

Mathis contends as well there is only the photon field so to speak interacting with electro-magnetism and gravity is the result. Now gravity is very peculiar in Mathis in that it is an "accelerator" pushing against the EM field on the quantum level and maintaining equilibrium at that atomic level. Gravity is a stronger force than QM currently believes at the atomic level and much weaker on the macro level than astronomers currently believe.

Am I correctly interpreting him? Let's let him speak for himself.

"I have solved this problem in the simplest way possible. I have shown that charge can be written as mass. In a dimensional analysis, charge and mass are equivalent.

To say it another way, charge must have mass. It already has mass in the defining equations. Quantum mechanics and quantum electrodynamics both rest on classical E/M theory and most of that theory has not even been tweaked, much less overthrown.

We only have to look at the definitions of
Coulomb, Ampere, and Tesla, to
realize that charge must have mass. The definitions
tell us that themselves, so it is incredible that QM
and QED ever forgot it

Once we are reminded that charge must have and
does have mass, these problems evaporate. Dark
matter is just charge photons. It is charge. It is not
something new, it is something very old, that even
Ben Franklin knew about."

Beyond this it is a matter of the balance between
the two forces and their relative strengths at the
macro and the micro level.

More clearly Mathis' definition of gravity is:

"To be specific, I will show that the gravitational
force is always a force in vector opposition to the
electromagnetic force, and that these two subtract
to give us a resultant force. This resultant force is
the one we measure and call gravity."

He elsewhere states that since gravity a derivative
of the EM force in opposition and that the EM
forces vary on the macro level from the micro
level that, therefore, the influence of gravity varies
between the two levels.

But that leaves us thirsting for the definition of the

generation of gravity itself. The above definition merely explains the strength of its force. That later.

Meantime I agree on the macro premise and the data appear to confirm that indeed gravity is the weaker force on the macro level.

On the micro-level I am not yet ready to judge. More later on this.

But that leaves us with the issue of the Higgs, does it exist on not. Or can we say that Mathis proposes a different kind of Higgs, a photon delivered Higgs. My head is now exploding. Does this make any sense at all? We'll see.

More tomorrow.

This will require some powerful thinking.

On Quantum Entanglement from Mathis

"Now let me show you some of the ways that entanglement is misinterpreted. If we go to Wikipedia for the modern gloss, we find this:

Quantum mechanics holds that states such as spin are indeterminate until such time as some physical intervention is made to measure the spin of the object in question. It is equally likely that any given particle will be observed to be spin-up as that it will be spin-down. Measuring any number of particles will result in an unpredictable series of measures that will tend more and more closely to half up and half down. However, if this experiment is done with entangled particles the results are quite different. When two members of an entangled pair are measured, one will always be spin-up and the other will be spin-down. The distance between the two particles is irrelevant.

A close reading of those few sentences already shows how the mystery of entanglement is a manufactured mystery, created by false probability assumptions. The problem in this Wiki quote is closely related to Schrodinger's cat mystery. In a thought problem, Schrodinger put a cat in a box and then assigned a probability number to the cat: say, .5 the cat was alive, .5 the cat was dead. We can't see the cat, so we don't know. Quantum mechanics says the numbers are all we know. Schrodinger says no, there is some fact underneath the numbers: either the cat is dead or it is alive. When we open the box, it must be one or the other, not both. Amazingly, the princes of QM did not say, "Yes, well of course. But we don't know until we open the box." That would have been sensible. No, QM said to Schrodinger that the cat was NOT really alive or dead. It was neither alive nor dead until we opened the box and saw it!

Yes, that is the level of philosophical

understanding of modern physicists. Schrodinger *lost* that argument, which is why I still have silly things like that that I can quote from Wiki."

One thing about Mathis, he doesn't shy away from the hard problems.

Here he is on entanglement:

"With entangled particles, "one will always be spin up and the other will be spin down." Note the word *always*. That is certain knowledge.

To explain this, quantum physicists have come up with the idea that the particles are in contact with each other over huge distances, without any mediating field or particle. Yes, they can talk to each other instantly, so that when the physicist measures one as spin up, the other can flip immediately to spin down to conserve parity.

All this is patently absurd, but neither the physicists nor the philosophers can seem to cut through to the fairly obvious answer."

Ok, I will bite, what is the answer about quantum entanglement?

8/1/12

Is entanglement absurd as Mathis
claims? Everything in Mathis' claims depends on
the nature and role of photons on the quantum
level.

Some researchers apparently agree so let's look at
what the researchers are finding about how the
photon interacts at the quantum level.

Here are some links and then we get back to
discuss.

http://www.sciencedaily.com/releases/2009/01/090
122141148.htm

Mathis argues for "stacked spins" in photons. What
does research show?

http://www.sciencedaily.com/releases/2010/02/100
214143131.htm

 But then there is this research which appears to
support the Mathis' view.
"Spooky action at a distance equals phony
quantum "magic?"

http://www.sciencedaily.com/releases/2011/06/110
624111942.htm

Entanglements can be multiple and galaxy wide?

http://www.sciencedaily.com/releases/2010/09/100

8/2/12 Spin or merely spin?

But first let's have a look at what Mathis means by spin.

http://milesmathis.com/wave.mov

Now let's have a look at a "stacked proton" with various spins.

Unifying the Electron and Proton

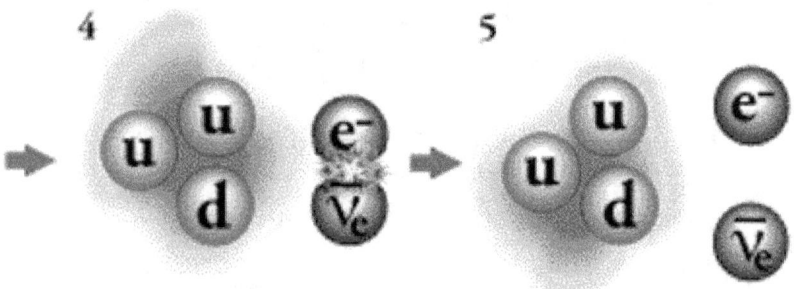

nope, standard model, wrong again

Abstract: Using simple math, I show that the electron is the proton stripped of its outer spins. See link below.

Finally Mathis says the "over-mystification" is what is happening here by the argument that "measurement affects electrons" which can be easily explained.

Here goes:

"Let's say you have a sample of electrons and are going to measure angular momentum in both zx and zy planes. If we have four possible outcomes, then we assume that each momentum is either clockwise or counterclockwise, relative to some observer. Now, put yourself in the position of this observer and see what happens. At the first moment, you look and you see that the electron is rotating clockwise about its x-axis, with that axis pointing straight at you. This means that the rotation is in the zy-plane. In other words, you are looking at a little clock, since it is moving relative to you just like the second hand on the face of a clock. That clock face exists in the zy-plane. A moment later the electron has rotated a half-turn, end over end along the x-axis. This rotation is in the zx-plane, about a traveling y-axis. After this half-turn, you look again at the clock face. Its motion is the same, but it now appears counterclockwise to you.

If that was confusing, you can easily perform

the above visualization with a desk clock, provided of course that it is not digital. Hold the clock in front of you. Its hands are turning clockwise, and they represent the spin in the x-plane. Now give the entire clock a spin in the y-plane, simply by turning it one half turn end over end. If you do this you will now be looking at the back of the clock. The second hand is now moving counterclockwise, relative to you. It is that simple. That is all I am saying. The second hand of the clock is spinning around an x-axis that is pointed right at you. Then you spun the whole clock around a y-axis. Very elementary, but it shows us that the x-spin of the electron must be variable, if you measure it relative to an observer external to the electron. If the electron has both x-spin and y-spin, then the x-spin will be variable, measured by a stationary device. Only an observer traveling with the electron would measure its spin as consistently CW or CCW. The same thing applies in reverse, of course. If you are measuring the other angular momentum, then you get a periodic variance in the first one."

Quote from:

http://milesmathis.com/super.html

So there you have it: Mathis' simpflied explanation. Does it really solve the problem?

Let's have a look next time...

But first a final look at superposition

http://milesmathis.com/super2.html

"Now let me show you how we can build the noble gases straight from alpha particles. But we also need the charge field. We must keep in mind that all these quantum particles exist in a charge field at all times. That is the second problem with mainstream theory. The first problem is that its diagramming is naïve. The second problem is that it doesn't know about the charge field. Photons are flying around in all directions, and this fact is important to our solution, as you will see.

http://milesmathis.com/nuclear.pdf

There is no nuclear strong force and why and the concept of holes, radiation and spin.

Mathis reconceives the periodic table showing the elements can be described as interactions of holes, the photon field, electrons and the rest. Here is an animation of Argon.

http://milesmathis.com/argon.avi

Mathis builds his ideas on the notion of the

physical existence of a "charge field." He uses it for everything.

See quote below:

"As a sort of postscript, I will point out to you that just as the charge field determines the structure of the nucleus; it also limits its size. We have seen that the protons and neutrons must position themselves to channel the charge field through and around the nucleus. This is done to prevent the charge field from pulling the nucleus apart. This limiting aspect of the charge field is what allowed me to understand the mechanics of nuclear structure, and it is ignorance of the charge field that had prevented nuclear diagramming before me.

Physicists have long known about charge, but they began hiding it in the math about 160 years ago (see Maxwell and quaternions). With QM and QED, charge went underground. It remained the defining and fundamental force of all the equations, but this foundation was purposefully obscured by the math.
Charge went from being physical to being mathematical to being wholly virtual. In the current equations it is nothing more than a ghost, which allows it to be ignored. As I have shown in many other papers, charge needed to be ignored to keep it from messing up the gravity field equations

of Einstein.

Physicists couldn't admit a charge field at the macro-level, because that would destroy all the field equations, all the way back to Laplace and Lagrange. That is why you find them misdirecting to this day. They have censored and slandered and built incredible walls to prevent this problem from seeing the light."

More on the charge field:

"Physicists couldn't admit a charge field at the macro-level, because that would destroy all the field equations, all the way back to Laplace and Lagrange. That is why you find them misdirecting to this day. They have censored and slandered and built incredible walls to prevent this problem from seeing the light. When charge is finally admitted to exist in the field equations, all their towers will crumble into dust.

But charge cannot be ignored or taken for granted. It must be given a physical and mechanical place in the field. I have shown in previous papers what was already known by Maxwell: charge has exactly the same notation as mass, and like energy it has a mass equivalence. It therefore cannot be virtual, cannot
be mediated by virtual photons, and cannot fail to take up space in the field. Photons have both

extension and mass equivalence. In fact, they turn out to have shocking amounts of both, and this fact has been buried in the electromagnetic field equations from the beginning. Basic equations going back to Maxwell prove that the charge field "outweighs" normal matter by 19 to 1, which means that dark matter is just charge. Most charge is dark, it is weakly interacting in the way dark matter physicists require, and it is much more massive that anyone knew. Ninety-five percent of the universe is not dark matter, it is photonic matter."

True or false I am beginning to see how the concept can be useful. But is it the Higgs, reformulated. Boy, there is real work here if we are going to get an answer. Hang in there for a bit longer. I think we are getting closer to being able to test for an answer. After all, if there is a real charge field we should be able to find it. Right?

Maybe we have and just have not recognized it?

But I leave you with this tantalizing quote: Tautological or brilliant? See what you think.

"The charge of the photon is the amount each photon is affected by other photons.
I mention this fact because it ties into my own

theory of missing mass. The missing mass isn't dark matter, it is the charge field. In other words, it is photons."

More later.

But first let's see what are the different kinds of photons Mathis is referring to and gain some insight there.

Meantime here is one definition of specific kinds of photons.

"Basic differences between the conventional and Aethermetric conceptions of the photon

1. On the nature of photons

1.1. Currently, it is held that solar radiation consists of photons. Implied in this is the notion that photons travel through space, like fibers of light, with analogy to ballistic models for the projection of material particles - as if the photons were hurled across space.

It is the view of Aethermetric theory that solar radiation does not consist of photons, but of the **mass free electrical charges** that compose the scalar electrical field [http://aetherometry.com/abs-AS2v2B.html#abstractAS2-17A]. Moreover, it is also the view of Aethermetric theory that photons are 'punctual' and local productions that they do not travel through space but rather occupy a globular space where they are created and extinguished.

1.2. If photons do not travel through space, what is it that travels through space and is the cause of the transmission of the light stimulus, and ultimately of any local production of photons?

Aetherometry contends that what travels through space and transmits the light impulse is electrical radiation composed of mass free charges and their associated longitudinal waves (the true phase waves), not electromagnetic radiation composed of photons and their transverse waves. The wave transmission of all electromagnetic signals depends on the transmission of nonelectromagnetic energy, specifically the transmission of electric mass free charges (the propagation of Òthe fieldÓ).

1.3. There are two types of photons: ionizing and nonionizing (blackbody). Aetherometry recognizes this accepted distinction, but suggests that it is a distinction still more profound than accepted physics itself holds, in that the two spectra are different as to the very conditions necessary for the production of one or the other type of photons. Specifically, Aetherometry claims that nonionizing or blackbody photons are locally generated whenever material particles that act as charge-carriers decelerate. Thus photons mark the trail of deceleration of mass bound particles. This punctual generation of photons that marks the trails of decelerating mass bound charges, combined with the decay in the kinetic energy of these charges, their release and scattered reabsorption by other adjacent mass bound charges (thus causing so called conversion of electromagnetic energy into longer wavelength radiation), is what accounts for (1) the dispersion of energy through conversion into electromagnetic radiation (and Tesla's persistent claim that his power transmitters were **not** transmitters of electromagnetic radiation) and for (2) the approximate suitability of the stochastic model for the dispersion of a ray and the scatter of light.

Conversely, material particles or mass bound charges accelerate when an electrical, magnetic, or electrical-cum-magnetic field is applied to them. Aetherometry contends that, in nature, an applied field is composed of mass free electric radiation, the effect of the radiation of mass free charges being the acquisition of their energy by the mass bound charges they encounter (*ergo* the addition of a kinetic energy term to the energy associated with the rest mass of a material particle), and thus the acceleration of these mass bound charges [http://aetherometry.com/abs-

AS2v2B.html#abstractAS2-16]. In summary, Aetherometry claims that 'radiation' of mass free charges is responsible for the acceleration of mass bound charges, whereas it is the **deceleration** of the latter which converts the lost kinetic energy into a local generation of blackbody photons."

More detail from Mathis on spin and the like.

http://milesmathis.com/stack.html
http://milesmathis.com/quark.html

Ok I am done for today.

More tomorrow.

8/4/12 Flux and Holes

Is there relationship between "flux tubes" and the "holes" of Mathis? Next time.

Meantime some reference material on flux tubes.

http://io9.com/5850729/quantum-locking-will-blow-your-mind--but-how-does-it-work

http://www.google.com/search?q=flux+tubes&hl=en&client=flock&hs=WaR&channel=fds&prmd=imvns&tbm=isch&tbo=u&source=univ&sa=X&ei=1YYdULj1GKbwiwK_3IHgAQ&ved=0CFkQsAQ&biw=1224&bih=498

http://en.wikipedia.org/wiki/Flux_tube

http://www.vets.ucar.edu/vg/MFT/index.shtml

http://blogs.discovermagazine.com/80beats/2011/10/20/superconductors-flux-tubes-frozen-levitating-puck/

Tesla's view of gravity?

http://www.youtube.com/watch?v=9Y0l0XELn1g&feature=related

We have a lot to ponder.

8/5/12

We now look squarely at Mathis' ideas on gravity and its propagation in conjunction with magnetic energy and his "charge field." We are getting close.

Magnetism, gravity as an accelerator, and of mice and frogs. Maybe Mathis' version of gravity has an explanation under superconductor conditions. There is oblique support for Mathis views on gravity by some authors and experimenters.

http://discovermagazine.com/1997/sep/floatingfrog1230/

http://blogs.discovermagazine.com/80beats/2009/09/10/two-things-you-need-to-levitate-a-mouse-1-strong-magnet-2-sedatives/

http://en.wikipedia.org/wiki/Diamagnetic

http://blogs.discovermagazine.com/80beats/2009/09/10/physicists-after-the-elusive-magnetic-monopole-spot-a-look-alike/

http://discovermagazine.com/2008/may/02-three-words-that-could-overthrow-physics/

"For one thing, as far as I can tell, *nobody knows* how a magnet can move a piece of metal without touching it. And for another—more astonishing still, perhaps—nobody seems to care.

This information was not easy to come by. My copy of *Electronics for Dummies* now shares a shelf with *Mathematics of Classical and Quantum Physics* by Frederick Byron Jr. and Robert Fuller. Should a doctor at any point take a cross section of my brain, she will find patches of scarring and dead tissue, souvenirs of the time I pursued the mystery of magnetism across the 11-dimensional badlands of string theory. Students of human pathos may one day cherish the 16-minute recording of me, with my 100 percent positive-feedback rating as an eBay purchaser, failing to make renowned physicist Steven Weinberg, who won a Nobel for unifying electromagnetism with the so-called weak force, admit that he can't explain how a magnet holds a dry-cleaning ticket to the door of a refrigerator.

But as far as I can tell—and isn't the point of science that all its bigger propositions come accompanied by this noble caveat?—he really can't"

More:

"When you get right down to it, the mystery of magnets interacting with each other at a distance has been explained in terms of virtual photons, incredibly small and unapologetically imaginary particles interacting with each other at a distance. As far as I can tell, these virtual particles are composed entirely of math and exist solely to fill otherwise embarrassing gaps in physics, such as the attraction and repulsion between magnets."

Quotes from the above article. Sounds like Mathis, doesn't it? Apparently the word about Physics' problems is on the street. This article was written in 2008.

Another example: Barbour is another critic sounding like Mathis.

http://discovermagazine.com/2012/mar/09-is-einsteins-greatest-work-wrong-didnt-go-far/article_view?b_start:int=1&-C=

"Barbour pressed on where Einstein had feared to tread, coming closer to Mach by dispensing not just with Newton's rigid grid but with the very concept of space-time. In general relativity, time is a dimension interwoven with the dimensions of space. In Barbour's universe, on the other hand, time is emergent: It is a measure of how space changes but not a fundamental component of it."

http://platonia.com/barbour_nature.pdf

Video that time like Mathis suggests only exists in specific circumstances

http://platonia.com/ideas.html#shape_dynamics

http://www.youtube.com/watch?v=WKsNraFxPwk&feature=player_detailpage

All of this at this point is giving me a headache. You?

It's Sunday. Let's take a walk.

8/6/12
More on Barbour and then we move on to biology and the human brain. What? I hear you say.

http://www.youtube.com/watch?v=z6wYhjyEfnI&feature=player_detailpage

http://www.youtube.com/watch?v=gOdk-u9wTkM&feature=player_detailpage

http://arxiv.org/pdf/gr-qc/0012089v3.pdf

http://www.youtube.com/watch?v=tJVDeusOfFM&feature=player_detailpage

Well if we are to understand atoms and time we have to also understand the atoms we know best, the ones which live in the human brain and in the human body.

Luckily I have done some of that homework already. Let's call it entanglement and the human brain and its mediums of communication.

It just may be that the theory of everything lies through the human brain and its physical connections. Hold on to your myelin sheath and we get back.

See the essays below for links to other essays. first and then, after, we get back and see what is in our Gigantic Bag of Speculations.

Einstein is Wrong Maybe-The Higgs Conference

Part Two

Einstein May Be Wrong: Part Two

Einstein and the Speed of Light Controversy

Einstein and the True Story of Relativity

Summary of Physics Theories

Einstein and the Nature of Space

Quantum Mechanics and the Speed of Light Controversy

Einstein, Time Space, Buddha and Literature

The Holographic View of the Universe

Einstein and a New Theory of How the Universe Works

Einstein: How Do Galaxies Work?

Einstein: The Plasma Theory of How the Universe Works